YOUR KNOWLEDGE HAS VALUE

- We will publish your bachelor's and master's thesis, essays and papers

- Your own eBook and book - sold worldwide in all relevant shops

- Earn money with each sale

Upload your text at www.GRIN.com and publish for free

Bibliographic information published by the German National Library:

The German National Library lists this publication in the National Bibliography; detailed bibliographic data are available on the Internet at http://dnb.dnb.de .

Imprint:

Copyright © 2014 GRIN Verlag, Open Publishing GmbH
Print and binding: Books on Demand GmbH, Norderstedt Germany
ISBN: 9783656601951

This book at GRIN:

http://www.grin.com/en/e-book/269167/the-schwarzschild-de-broglie-modification-of-special-relativity-for-massive

Siegfried Gantert

The Schwarzschild-de Broglie Modification of Special Relativity for Massive Field Bosons (SBM)

A study about dark matter and dark energy from the SBM model perspective

GRIN Publishing

Scientific Study

The Schwarzschild-de Broglie Modification of Special Relativity for Massive Field Bosons (SBM)

A study about dark matter and dark energy from the SBM model perspective

Siegfried Gantert

15.02.2014

Table of Contents:

Abstracts   1

1.0   Introduction   1

2.0   The Schwarzschild-de Broglie modification of SRT   2
for massive field bosons (SBM)

3.0   Spontaneous symmetry breaking with the formation   8
of a phase boundary

4.0   Higgs mechanism from the SBM model perspective   13

5.0   Discussion and conclusion   17

6.0   Acknowledgement   19

7.0   Literature   19

The Schwarzschild-de Broglie Modification of Special Relativity
for Massive Field Bosons (SBM)

A study about dark matter and dark energy from the SBM model perspective

Author: Siegfried Gantert
s.gantert@email.de

Abstract

This work is a presentation of a modified form of special relativity for field-bosons – in short SBM. Field bosons, for the purposes of this work, are synonymous with the condensates from spin 0-particles. The starting point is the hypothesis that a minimum size of uncertainty $\Delta x > 2 \cdot r_S$ (r_S=Schwarzschild radius) becomes effective with relativistic velocities, from which different limit velocities $0 < v_{l,n} < c$ are derived, depending on the size of the field bosons. In accordance with the SBM model, field bosons under a defined phase limit become massive through spontaneous symmetry breaking. Field bosons can melt into larger condensates through the effects of gravity, whereby their effective mass is reduced, thus also reducing their large-scale gravitative coherence.

1.0 Introduction

According to the latest background radiation measurements (CMB) of the ESA's Planck satellite, about 68 percent of the universe is composed of a mysterious dark energy and 27 percent of dark matter, with the more familiar visible matter only making up about 5 percent of the universe [1].

F. Zwicky was one of the first to point out the possible existence of dark matter, after investigations into the proper motion of galaxies in galaxy clusters showed great compatibility with this idea [2, 3]. Similar observations were made in the rotation curve of spiral galaxies [4-9]. In these cases, the amount of dark matter increases with distance from the center, while it is comparatively low in the center of the galaxy.

However, the result of an investigation into the distribution of mass in the matter around the sun [10] showed that this mass is distributed almost exactly as it appears visibly. There were thus no indications for dark matter found.

Due to the difficulties of reconciling these sometimes-contradictory observations with known physical concepts, many concepts of dark matter and dark energy have been proposed. Some models that have attempted to describe the enigmatic behavior are string theory [11], loop quantum gravity [12], the quintessence model [13], the axion [14], phantom energy [15], and the MOND theory [16].

If one takes the cosmological constant Λ as a basis, the background radiation data [17, 18] regarding the distribution of dark matter and dark energy can be well reconciled with the predictions of the cosmological standard model (Λ-CDM model). However, the fundamental mechanisms that could explain a dark energy that works against gravity remain largely unexplained at this point.

The predictions of the Λ-CDM model also lead to inconsistencies in order of magnitude scales of a galaxy [19]. In terms of the model's projections, the center of a galaxy should rotate faster than the observed measurements. One would also expect to observe a greater density of cold dark matter towards the center of a galaxy. The discrepancy between the measurements and the predictions of the Λ-CDM model were referred to as a "cold dark matter catastrophe" in astrophysics literature [20].

Mayer et al. ascribed the lack of dark matter in the center region of a galaxy to supernova explosions [21]. A research group led by Benoit Famaey came to the conclusion, on the basis of their investigations [22, 23], that there must be a close connection between the distribution of visible matter and dark matter. Their observations correspond more with the predictions of the MOND theory (Modified Newtonian Dynamics) of Mordehai Milgrom [24].

Models interpreting dark matter as a condensate of a scalar boson field have recently received increased attention [25-38].

At the elementary particle level, hypothetical particles such as the axion or WIMPs (weakly interacting massive particles) have been considered as candidates for dark matter. In the SUSY theory(-ies), it is the neutralino, the lightest (LSP) of a series of hypothetical particles, which is seen as a possible candidate for cold dark matter (CDM) [39].

2.0 The Schwarzschild-de Broglie modification of SRT for massive field bosons (SBM)

The work presented here aims to utilize astrophysical data [1-10, 50] in order to check the results of the SBM model for consistency. It is motivated by the possibility of better understanding the physical aspects of dark energy and dark matter.

The work is structured as follows: the underlying SBM model will first be presented along with the different limit velocities of field bosons with masses of various scales of magnitude derived from the model. Subsequently, the phase limit will be determined at which the relativistic field bosons receive their effective mass through spontaneous symmetry breaking. In the following section, a mechanism similar to Higgs will be proposed, on the basis of the SBM model, in order to place the total energy of the field bosons involved into a quantitative relationship with their mass-giving effect. In conclusion, the results of the findings will be employed to discuss the usefulness of the proposed SBM model, using examples from some astrophysical problem areas.

Within the concept of the Planck scale, the Planck length is often taken as a limit for the validity of currently known physical laws [40]. The Planck length (3) is determined through a comparison of the size of the Schwarzschild radius (1) with the Compton wavelength, equation (2). According to a concept from Max Planck [41, 42], the description of physical phenomena over distances smaller than the Planck length (1.616E − 35m), equation (3), is impossible with our current level of scientific understanding, and can only be formulated with a quantum theory of gravitation.

$$r_s = {2 \cdot G \cdot m}/{c^2} \qquad\qquad \text{Schwarzschildradius} \qquad (1)$$

$$\lambda_C = \frac{h}{m \cdot c} \qquad\qquad \text{Comptonwavelength} \qquad (2)$$

$$l_P = \sqrt{\frac{\hbar \cdot G}{c^3}} \qquad\qquad \text{Planck length} \qquad (3)$$

In principle, the Compton wavelength can be determined by the increase in wavelength of the scattered radiation at a right angle that results from an elastic impact of a photon with a particle at rest. After the collision, this normally leads to the suppression of the coherent properties of quantum mechanical states, which is related to the loss of the interference capability of the particle. As photons do not collide with dark matter, having only a gravitational interaction, it is probable that a stable and coherent system exists. It thus seems worthwhile to examine the appearance of dark matter in terms of a coherent interaction of bosons, especially the coherent interaction of Higgs bosons.

Due to their integral spins, Higgs bosons can arrange themselves into condensates, given stable quantum mechanical conditions. It can be shown that appropriate potentials develop with a certain modification of the special relativity theory, and that these could contribute to the development of such condensates. In the context of this work, the term "field boson" will usually be used in place of condensate in order to highlight the particulate character of such an assembly. Because Higgs bosons, as the constituents of such assemblies, are subject to Bose-Einstein statistics due to their integral spins, a respective field boson (condensate) can be described as a single particle wave function and can be understood quantum mechanically as an independent particle.

Therefore, the Compton wavelength, as a collision result, will not be taken as the central point of consideration in this work, but rather the de Broglie wavelength (4) of field bosons of differing sizes.

$$\lambda_{dB} = h/p \qquad\qquad \text{De Broglie wavelength} \qquad (4)$$

The SBM model presented here differs not only from the SRT of Albert Einstein in that it differentiates between fermionic and bosonic particles; it also differs from the double special relativity theory [43-45], which takes the approach that Planck length (or Planck energy) should be taken into account, in addition to the speed of light, as a further Lorentz invariant quantity.

First, possible consequences of coherent conditions of Higgs bosons with regard to relativistic speed behavior of field bosons will be discussed.

As it is impossible to make any remarks about the strength of the particles' interconnections a priori, the scale mass of a field boson is defined as the sum of the masses of the (Higgs) bosons of which it is composed, equation (5).

$$\sum_{n=1}^{n} m_H = n * m_H = M_n \qquad\qquad \text{Scale mass } M_n \qquad (5)$$

The scale mass should mainly serve as a reference scale to show the effective sizes of the field bosons.

For further approaches, the special relativity theory (SRT) was used, with the reservation that a description of a relativistic particle of the field particle type described above is impossible in the SRT. As a result, the laws of SRT were only used to lead towards the point where a "collapse" in wave function was expected. In a second step, a specific relativistic limit velocity $v_{l,n} < c$ that prevented a collapse in wave function, or the loss of coherence was searched for every field boson. A "l in the index of limit velocities $v_{l,n}$ stands for "limit," while "n" designates the number of Higgs bosons constituting a specific field boson. If there are no attractive or repellant interactions between the Higgs bosons, equation (6) applies to the de Broglie wavelength of a (free) field boson with mass center M_n, where p_n stands for the impulse of the field boson.

$$\lambda_{dB,n} = \frac{h}{p_n} \qquad \text{De Broglie wavelength, field boson} \qquad (6)$$

If the field boson moves very quickly from the point of view of an observer at rest (inertial system), then the relativistic impulse p_n in equation (7) must be taken into account.

$$p_n = \frac{M_n \cdot v}{\sqrt{1 - (v/c)^2}} \qquad (7)$$

From the point of view of an observer at rest ($v = 0$) the matter wave of a coherent field particle is compressed in the direction of motion at relativistic speeds v. It is now assumed that, for an observer at rest, the material wave around the limit speed $v_{l,n}$ of a field boson passes over the stage of a flattened rotation ellipsoid into a toroidal limit form; it is simplified into a horn torus, as shown in Fig. 1. For the uncertainty Δx of such a field particle close to the speed limit, there is a minimum size, with relation (8), of which the value is larger than the double of the Schwarzschild radius.

$$\Delta x > 2 \cdot r_s \qquad (8)$$

The validity of the relativistic length contraction per (9) close to the mass center can actually be sustained with requirement (8), without it leading to a collapse of the matter wave or the creation of a punctiform singularity, see Fig. 1.

$$L_n = L_{0,n} \cdot \sqrt{1 - \frac{v^2}{v_{l,n}^2}} \qquad (9)$$

From the point of view of an observer at rest, the "length" of a field boson under limiting condition (8) approaches 0 at relativistic energies at the mass center, as shown in Fig. 1. The de Broglie wavelength relationship (10) applies to this limiting condition:

$$\lambda_{dB} \cong 2 \cdot r_s \cdot \pi \qquad (10)$$

The dotted lines at the right in Fig. 1 schematically show the outer Schwarzschild solution as a sectional drawing through a Flamm's paraboloid in two different position with the distance Δx. In the SBM model, this thus results in differences with the SRT, for which a break in known physical laws is predicted near the Planck length [41].

Figure 1: Toroidal symmetry of the limit configuration (horn torus) with the de Broglie wavelength $\lambda_{dB} \cong 2 \cdot r_s \cdot \pi$ for determining the relativistic limit velocity $v_{l,n}$ of a field boson (sectional drawing at right).

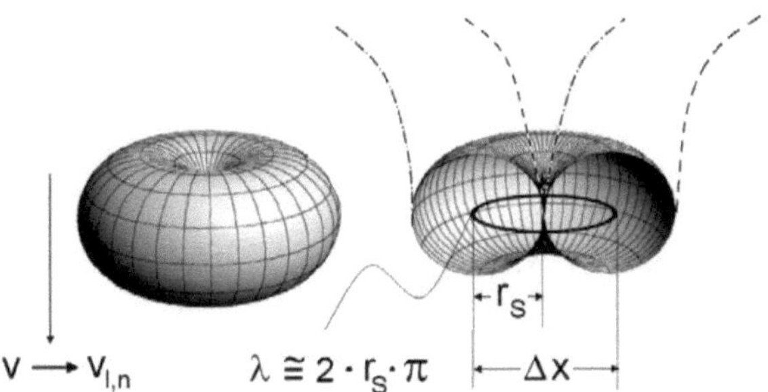

For the explicit calculation of the limit velocity $v_{l,n}$ of a field boson of scale mass M_n, relationship (10) will be used to establish the conditional equation (11) for $v_{l,n}$, taking into account the Lorentz factor γ.

$$\frac{\hbar \cdot \sqrt{1 - \frac{v_{l,n}^2}{c^2}}}{M_n \cdot v_{l,n}} = \frac{2 \cdot G \cdot M_n}{c^2 \cdot \sqrt{1 - \frac{v_{l,n}^2}{c^2}}} \qquad \text{Limit case } v = v_{l,n} \text{ to determine } v_{l,n} \quad (11)$$

Solving the equation (11) for $v_{l,n}$ gives (12) as a solution, whereby only positive values should be taken into account before the radical term.

$$v_{l,n} = -\frac{G \cdot M_n^2}{\hbar} \pm \sqrt{\frac{G^2 \cdot M_n^4}{\hbar^2} + c^2} \qquad (12)$$

The graphical presentation of the limit velocities $v_{l,n}$ (12) contrasted with the scale mass M_n (Fig. 2) shows that below a scale mass of about 1E-10 kg there are hardly any noticeable deviations of the SBM from the special relativity theory (SRT), as applies $v_{l,n} \cong c$ to field bosons with a low scale mass.

However, above 1E-10 kg on the mass scale, there are stronger deviations from the SRT. The calculated limit velocities gradually approach 0 as scale masses increase, see Fig. 2.

According to equation (12), the limit velocities $v_{l,n}$ are only dependent on the scale mass M_n (3) of a field boson and the natural constants G, $\hbar$ and c. In the case of a bosonic particle without resting mass, a photon for example, applies $v_{l,n} = c$, as one can easily verify from equation (12) for $M_n = 0$. As will be shown in more detail later, contrary to the SRT, the SBM predicts the dispersion of limit velocities of field bosons with non-zero resting masses. Bosons without resting mass, for example photons (of differing energies), should therefore not display any velocity dispersion. This conforms well to run time measurements of photons from GBR events [46, 47] (GBR gamma ray burst).

Figure 2: Calculated limit velocities $v_{l,n}$ according to the SBM model on a mass scale from 0 to about 1.6E-7 kg.

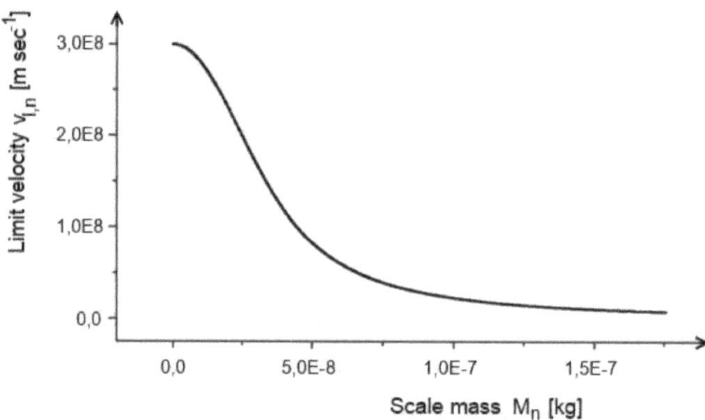

If one retains relativistic principles and substitutes the newly acquired limit velocities $v_{l,n}$ for the speed of light c in the energy impulse equation of the SRT (13), or in equation (15) for total energy, there is a strong deviation from the SBM model, in contrast to the SRT, see Fig. 3 below.

$$E_{SRT,n}^2 = E_0^2 + (c \cdot p_n)^2 \quad (13) \qquad E_{SBM,n}^2 = E_0^2 + (v_{l,n} \cdot p_n)^2 \quad (14)$$

$$E_{SRT,n} = \frac{M_n \cdot c^2}{\sqrt{1-\frac{v^2}{c^2}}} \quad (15) \qquad E_{SBM,n} = \frac{M_n \cdot v_{l,n}^2}{\sqrt{1-\frac{v^2}{v_{l,n}^2}}} \quad (16)$$

With the introduction of the Schwarzschild – de Broglie modification of the SRT, regarding (14) and (16), there is a strong decrease in resting energy ($v = 0$) with increasing scale mass of the field bosons, see Fig. 3 below. The numbers given on the curve progression correspond to the scale mass of the depicted Schwarzschild de Broglie condensate (field boson) in kg.

In comparison to this, Fig. 3, above, shows the theoretical progression of the total energy of particles of differing heaviness according to the SRT. The resting energies of the particles ($v = 0$) behave according to the equivalence of mass and energy, while, in the case of field bosons (see Fig. 3 below), they decrease as their scale masses gradually increase. The maximum resting energy is achieved at about 0.597E-9 J with a scale mass of 1.107E-8 kg, whereby the Planck energy is significantly higher at 1.9561E9 J.

As the equivalence of scale mass and energy within the SBM model can no longer be supported (see Fig. 3), a differentiation will now be made between scale mass and the effective mass of a field boson.

Figure 3: Above: velocity – energy plot of different particle masses according to Einstein's SRT, and below, in comparison, field bosons of differing scale masses according to the SBM model. The number given on the curve in Fig. 3 (below) corresponds to the scale mass of the depicted Schwarzschild-de Broglie condensate (field bosons), in kg.

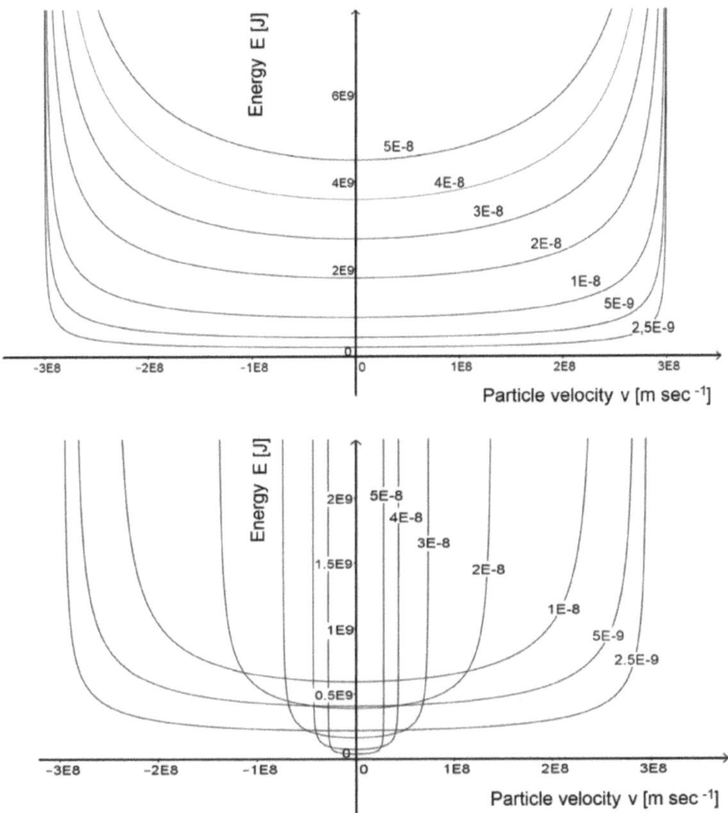

In Fig. 4, the scale mass is given on the x-axis against the total energy of the field bosons at different velocities. The effective mass of a field boson at a certain velocity is determined by its energy value (see text below). The diagonal dotted lines in Fig. 4 show the course of the resting energy at comparable scale masses in accordance with the SRT. As one can see, the velocity trajectories are found in a further scale mass area below the energies according to the SRT version.

Figure 4: Diagram of scale mass vs. total energy of field bosons of different velocities, according to the SBM model. The numbers on the parameter curves give the velocity in units of the speed of light.

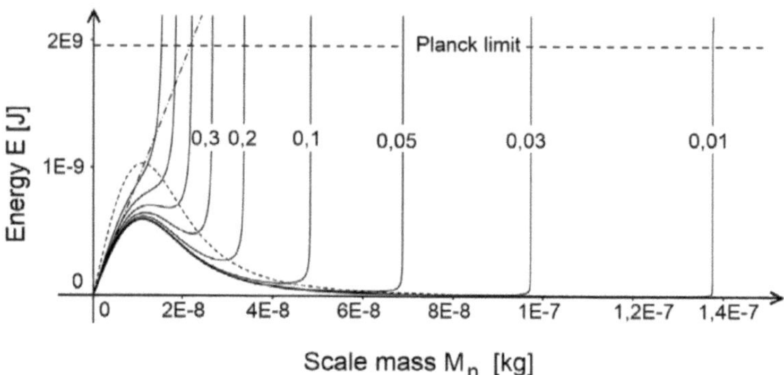

To create a relationship between the scale mass of a field boson with its effective mass, the measurement μ_n (17) is introduced, which reflects the relation of the limit velocity of a field boson to the speed of light.

$$\mu_n = \frac{v_{l,n}}{c} \tag{17}$$

With definition equation (17), equation (16) of the SBM for $v = 0$ may now be reformulated in the form of (18) as well,

$$E_{n,SBM} = M_n \cdot \mu_n^2 \cdot c^2 \tag{18}$$

so that the expression (19) for the effective (invariant) mass $M_{eff,n}$ results from the division by c^2.

$$M_{eff,n} = M_n \cdot \mu_n^2 \tag{19}$$

Fig. 4 also includes a dotted curve that corresponds to equation (20), which for large scale masses shows a good approximation for the calculation of the position of the minimums of the various parameter curves shown in Fig. 4.

$$E_{min} \cong M_n \cdot v_{l,n}^2 \cdot \sqrt{3} \quad \text{for large scale masses} \tag{20}$$

3.0 Spontaneous symmetry breaking with the formation of a phase boundary

In Fig. 5, the approach to date for determining the limit velocity $v_{l,n}$ is again shown graphically in the diagram with the example of a field particle with the scale mass M_n. With the help of the SRT and hypothesis (8), the limit velocity $v_{l,n}$ can be determined at point A, at which point the SB modification of the SRT can be performed in line with equation (16). The result of the modification can be seen in Fig. 5 along the new SBM curve.

Figure 5: An explanation of the Schwarzschild de Broglie modification of the SRT for a field boson with the scale mass M_n: spontaneous symmetry breakdown at point B (see text for further information).

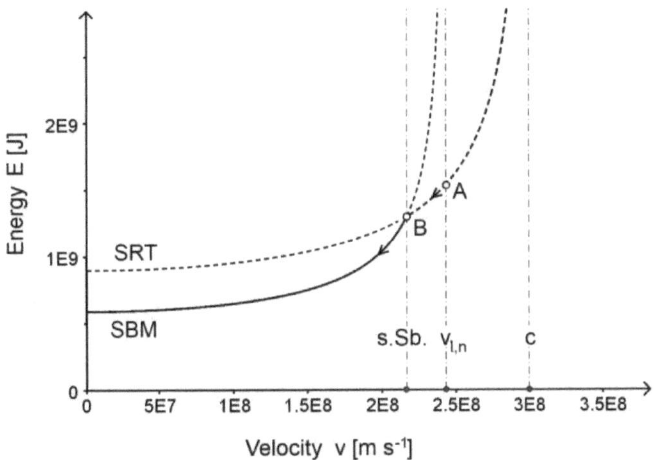

The original SRT curve crosses the SBM curve at point B. Intersection B is the part of a phase boundary line where the transition into the lower SBM phase under spontaneous symmetry breakdown occurs. A massive field boson with toroidal symmetry is thus created (see also Fig. 1). As the exit constituents at point B must be very close together at the creation of field bosons, conditions are necessary that probably only existed shortly after the Big Bang. However, Fig. 5 and all following plots and diagrams show energetic conditions in flat space, i.e. without the influence of gravitation. Under these circumstances, a field boson decays irreversibly, in the opposite direction, during a phase transition.

The velocity of a field boson $v_{ph,n}$ (21) at the phase boundary can be determined by equating (15) with (16). See Fig. 5.

$$v_{ph,n} = \sqrt{\left.(v_{l,n}^4 \cdot c^2 + v_{l,n}^2 \cdot c^4)\right/(v_{l,n}^4 + v_{l,n}^2 \cdot c^2 + c^4)} \tag{21}$$

The situation can be made somewhat clearer if one uses the quotient $\Phi(M_n, v)$, equation (22), from which the energy is modeled as a function of the scale mass, according to the SRT and SBM models. See Fig. 6:

$$\Phi(M_n, v) = \left.E_{SBM}\right/E_{SRT} = \mu^3 \cdot \sqrt{\left.(c^2 - v^2)\right/(v_{l,n}^2 - v^2)} \tag{22}$$

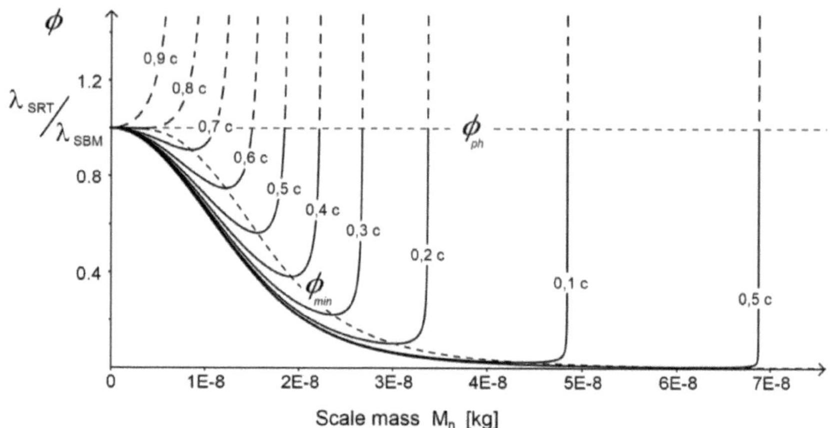

For the phase boundary line Φ_{ph}, below which the field bosons become massive, applies
$E_{SRT} = E_{SBM}$. This is why it follows exactly $\Phi_{ph} = 1$ in Fig. 6 – see also the conditions for
intersection point B in Fig. 5. The phase boundary line Φ_{ph} , equation (23), can also be found
by substituting v with $v_{ph,n}$ in equation (22). However, the values Φ_{ph} do begin to fluctuate
at large scale masses due to insufficient accuracy.

$$\Phi_{ph} = \mu^3 \cdot \sqrt{\left(c^2 - v_{ph,n}^2\right) \Big/ \left(v_{l,n}^2 - v_{ph,n}^2\right)} = 1 \tag{23}$$

The progression of the dotted curve Φ_{min} in Fig. 6 passes through all minimums of the
parameter curves $\Phi(M_n, v)$ and can be graphically shown using equation (24).

$$\Phi_{min} = \mu^3 \cdot \sqrt{\left(c^2 - \tfrac{2}{3} v_{l,n}^2\right) \Big/ \left(v_{l,n}^2 - \tfrac{2}{3} v_{l,n}^2\right)} \tag{24}$$

According to the Hamilton principle, field bosons can condense along the curve Φ_{min} in Fig.
6 into field bosons with greater scale mass, in order to reach lower potentials. The driving
force for this is the inner deformation pressure of the coherent Higgs boson at higher energies
or velocities on the field bosons. As field bosons with slower velocities and lower effective
masses are created at a related temporal development – along the curve Φ_{min} in Fig. 6 – the
result is a decrease in gravitational coherence. From the point of view of the SBM model, a
slow and sustained inflation [48, 49] thus seems plausible, even though the potential curves in
Fig. 6 can only show the simplified conditions in an asymptotic flat space-time. A part of dark

energy can thus be understood as a result of a condensate of field bosons, which possibly accompany the clumping of baryonic material within the galaxies.

The parameter curves in Fig. 6 can also be interpreted in two phases in terms of the de Broglie wavelength conditions. Equation (25) applies to the effective (relativistic) de Broglie wavelengths according to the SBM:

$$\lambda_{SBM} = \left. h \cdot \sqrt{1 - (v/v_{l,n})^2} \middle/ M_{eff,n} \cdot v \right.$$
(25)

As is plain to see here, the quotient E_{SBM}/E_{SRT}, equation (22), correlates to the reciprocal relationship of both de Broglie wavelengths according to the SRT and SBM representations of field bosons, see equation (26).

$$\Phi(m_n, v) = E_{SBM}/E_{SRT} = \lambda_{SRT}/\lambda_{SBM}$$
(26)

Therefore, the effective de Broglie wavelength λ_{SBM} according to the SBM needs to be at least as large as the de Broglie wavelength λ_{SRT}[1] , so that field bosons acquire mass through symmetry breaking or fermionic elementary particles are transferred (invariant) mass by massive field bosons, see Fig. 6. In Fig. 7 the different velocities of parameterized curves of the field bosons per equation (16) are shown with the respective phase boundary line (27) in the total energy E vs. scale mass diagram.

$$E_{SBM,ph} = \left. M_n \cdot v_{l,n}^2 \middle/ \sqrt{1 - \frac{v_{ph}^2}{v_{l,n}^2}} \right.$$
(27)

The vertical numbers in Fig. 7 again stand for the velocity of field bosons in units of speed of light. The dotted diagonal (A) in the diagram shows an example of the total energy of field bosons with increasing scale mass M_n at $v = 0{,}6 \cdot c$ according to the SRT. It intersects the deviating SBM curve (B) just at the phase boundary line, beneath which the SBM system for massive field bosons begins.

---

[1] i.e. mit $v_{l,n} = c$, see equation (6) und (7)

Figure 7: Scale mass vs. total energy (M_n-E plot) with phase boundary line

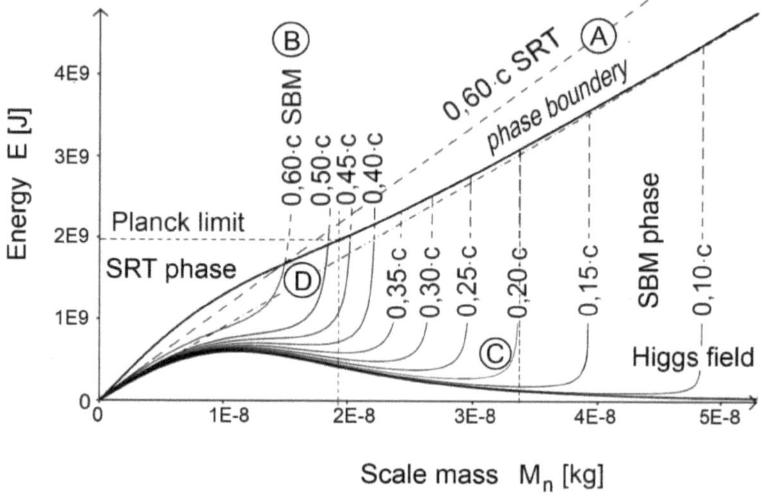

The diagonal line D, which asymptotically approaches the phase boundary line in Fig. 7, corresponds to the resting energy curve of the related scale masses per the SRT version. The Planck limit curve at E=1.96751E9 J is also shown in Fig. 7. Until a scale mass of about 1.9E-8 kg, there are two phase boundaries below the Planck limit in which the SRT system predominates at higher energies and the SBM system at lower energies. According to Fig. 7, field bosons with a scale mass larger than 1.9E-8kg can at least theoretically exist above the Planck limit until the SBM/SRT phase boundary.

Above a scale mass of about 1.5 E-8kg, or under a velocity of about 0.45 c, all curves of constant velocity traverse an energy minimum before they rise steeply to the right at a certain scale mass. As an example, the velocity trajectory of $0,2 \cdot c$ (relative to an observer) with a scale mass of about 3E-8 kg corresponds to the energy minimum (C), while the related vertical asymptote intersects the x-axis at 3.372 E-8 kg. Therefore, the field boson with a scale mass of 3.372 E-8 kg has the limit velocity of $0,2 \cdot c$.

According to the Hamilton principle, an observer from a distant spiral galaxy would have to reckon with field bosons of differing effective mass $M_{eff,n}$, depending on the rotation speed of the galaxy disc in relation to him. If one considers, for example, the position of the minimum energy of certain velocity trajectories in Fig. 7 relative to the scale mass, and compares them to different rotation velocities within a galaxy disc, one would expect field bosons with larger scale mass M_n, but smaller effective mass $M_{eff,n}$, when close to the galaxy center with a relatively low rotation velocity. In the case of a higher rotation velocity and increasing distance from the galaxy center, one would expect field bosons with a lower scale mass, but with higher effective mass. This qualitatively illustrates a view that is compatible with the distribution of so-called dark matter within spiral galaxies [4-9], and it additionally offers a simple explanation for the Core-Cusp problem [50].

4.0 Higgs mechanism from the SBM model perspective

In order to be able to make quantitative statements about the problem of dark matter or dark energy from the SBM model perspective, the effect of field bosons on a series of "massless" fermionic test particles of a specific energy will be examined at the limit case of a flat space-time. The goal is to set the total energy of the field bosons involved into a quantitative relationship with their mass-transferring effects, in order to later be able to use this as a model for dark matter and dark energy in an asymptotic flat space-time.

According to the SBM, field bosons "recognize" a fermionic test particle (elementary particle) by the size of the quotients of the equation (22) or the proportional relationships of both de Broglie wavelengths, equation (26). If the effective de Broglie wavelength, equation (25), of a massive field boson is greater than the de Broglie wavelength of the test particle, then the correspondent field boson can contribute a quantity to the mass. Fig. 8, left, demonstrates the connection between the energy of a test particle and that from a specific resultant projection mass $M_{p,n}$. It delineates the lower limit of the mass scale, above which the field bosons can contribute mass to the test particle. As will be demonstrated in detail, one requires $M_{p,n}$ as a scale factor (28) to form a velocity or energy distribution of the mass-transferring field bosons.

$$SF_{SBM} = M_{p,n} \cdot c^2 \qquad \text{Scale factor for distribution functions (28)}$$

As shown in Fig. 8, at left, point $P(M_{p,n}/E_T)$ can be constructed as the intersection of the energy E_T of the test particle with the phase boundary (Fig. 8, phase boundary), equation (23). The related projection mass $M_{p,n}$ is then found from the intersection of the perpendicular through point $P(M_{p,n}/E_T)$ with the mass scale.

Figure 8: Higgs mechanism from the perspective of the SBM:

Left: Connection between the energy of a test particle and the projection mass $M_{p,n}$.

Right: Possible field boson transfers (selection) to a test particle of energy E_T, with projection mass $M_{p,n}$ (T=test particle, p=projection).

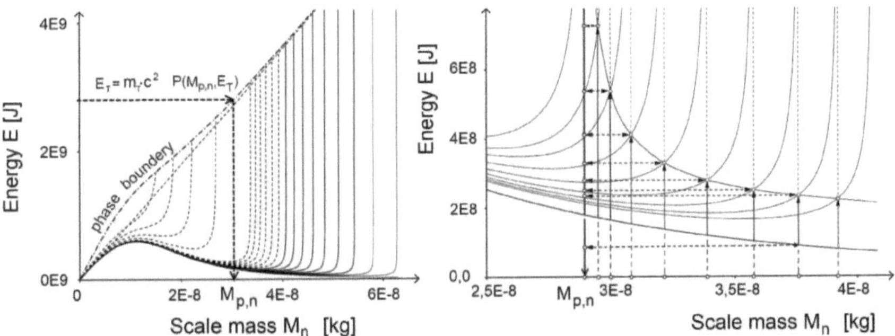

To the right of the connection line P(M$_{p,n}$/E$_T$) - P(M$_{p,n}$/0) is the phase area in which field bosons can deliver an amount of mass to the test particle. This is a result of the described interaction of the de Broglie wavelength conditions at the phase boundary (see the explanations of equations (22) to (26)). Field bosons that have effective de Broglie wavelengths (25) smaller than the de Broglie wavelengths of the test particles cannot contribute any quantity to the mass of the test particle. These are found in Fig. 8 left of the connection line P(M$_{p,n}$/E$_T$) - P(M$_{p,n}$/0).With regard to a test particle, the field boson with the projection mass $M_{p,n}$ in particular is a virtual particle due to the infinitesimal probability of transference, see also Fig. 8, right.

The relativistic velocity or energy distribution of the virtual field boson with a scale mass $M_{p,n}$ of now represents a single particle projection of the mass-transferring field boson transitions right of the connection line P($M_{p,n}$/E$_T$) - P($M_{p,n}$/0) (in Fig. 8, at right). These transitions are only of a very short duration, as the deformation pressure of the coherent Higgs bosons forces a field boson back to the energy level of the Higgs field ($v = 0$).

As the questionable transitions are relativistic due to the comparatively small limit velocities of the field bosons, it is here suggested to present the single particle projection through a modified Maxwell-Jüttner velocity distribution (MJV) corresponding to equation (30) for simplicity from the point of view of a co-moving observer. The original scale factor kT of the Maxwell-Jüttner velocity distribution (MJV) (29) is replaced by the energy equivalent of the projection mass $M_{p,n}c^2$ in the modified MJV (30).

Table 1: Contrast of the modified MJV (30) for massive field bosons with the MJV (29)

Maxwell Jüttner Distribution	Modified Maxwell Jüttner Distribution

$$f(\gamma) = \frac{\gamma^2 \cdot \beta}{\theta \cdot K_2(1/\theta)} \cdot \exp(-\frac{\gamma}{\theta}) \quad (29)$$

$$f(\gamma_p) = \frac{\gamma_p^2 \cdot \beta_p}{\theta_\kappa \cdot K_2(1/\theta_\kappa)} \cdot \exp(-\frac{\gamma_p}{\theta_\kappa}) \quad (30)$$

with

with

$$\gamma = \frac{1}{\sqrt{1 - \dfrac{v^2}{c^2}}} \quad (31a)$$

$$\gamma_p = \frac{1}{\sqrt{1 - \dfrac{v^2}{v_{p,l}^2}}} \quad (31b)$$

$$\mu_p = \frac{v_{p,l}}{c} \quad (32)$$

$$\theta = {kT}/{m \cdot c^2} \quad (33a)$$

$$\theta_\kappa = {\kappa \cdot M_{p,n} \cdot c^2}/{M_{p,n} \cdot \mu_p^2 \cdot c^2} = {\kappa}/{\mu_p^2} \quad (33b)$$

$$\beta = \frac{v}{c} = \sqrt{1 - {1}/{\gamma^2}} \quad (34a)$$

$$\beta_p = \frac{v}{v_{p,l}} = \sqrt{1 - {1}/{\gamma_p^2}} \quad (34b)$$

The MJV in the form of equation (29) gives the probability that a particle with mass m will take a specified γ- value at a given temperature, γ, θ and β are defined in the MJV through the equations (31a), (33a), and (34a). $K_2(1/\theta)$ is a modified Bessel function of second kind.

In the MJV (29), the mass of the relevant particle is already an integral part of the distribution function and has a formulary relationship with temperature; see sub-equation (33a). The requirement for such a linkage is however first present when the test particle already has mass and can interact with an electromagnetic field, for example as a fermion. The latter is known to be an atypical property of dark matter. In comparison, in the modified MJV (30), the projection mass $M_{p,n}$ of a virtual field boson can be determined alone by the position of the test particle relative to the phase limit curve; it thus does not represent any anticipated assumption in terms of the mass m_T of the test particle.

In place of the mean thermal energy kT, the modified MJV (30) introduced the energy equivalent E_T of the projection mass $M_{p,n} \cdot c^2$ derived from the particle energy as a temperature-independent scale factor. The position of $M_{p,n}$ on the mass scale ensures that the projections of the field boson excitement can be shown completely on the connection line between points $P(M_{p,n}, E_T)$ and $P(M_{p,n}, 0)$ in Fig. 8. Due to the curve of the phase boundary (see Fig. 8 at right), $M_{p,n}$ deviates from the mass m_T of the test particle. With the help of the calculated probabilities $f(\gamma_p)$ of the modified MJV, per equation (30), the total energy E_g of all participating mass-transferring field bosons can be determined through the integral (35).

$$E_g = \int_{\gamma_p=1}^{\gamma_p=\infty} E(\gamma_p) \cdot f(\gamma_p) \, d\gamma_p \tag{35}$$

In equation (35), $E(\gamma_p)$ stands for the energy of the virtual field boson with scale mass $M_{p,n}$, with Lorentz factor γ_p, per equation (36):

$$E(\gamma_p) = \frac{M_{p,n} \cdot v_{p,l}^2}{\sqrt{1 - \dfrac{v^2}{v_{p,l}^2}}} \tag{36}$$

The integral (35) is numerically determined and evaluated on the basis of $6 \cdot 10^4 \ \gamma_p$ - values for a series of test particles of various energies.

Additionally, the linkage parameter κ in θ_κ, sub-equation (33b), was introduced as a helpful expansion of the modified Maxwell Jüttner distribution. This allows the proportion of the total energy (35) of the field bosons to be identified, which stands for the generation of the mass of the test particle (37). The geometrical meaning of the linkage parameter κ will be further addressed below. In Fig. 9 at left, the probability distribution $f(\gamma_p)$, equation (30) is shown for the κ- values 1/3, 2/3 and 1 for $M_p = 2.895E\text{-}8$ kg, i.e., for a test particle energy E_T of $2.69627E\text{+}09$ J. Fig. 9 at right shows the correspondent energy distribution of the field bosons in the single particle representation, with different γ_p values for the κ- values 1/3, 2/3 and 1.

Figure 9: Probability distribution (left) and energy distribution (right) vs. γ_p-values at different κ-values in the single particle representation of the modified Maxwell Jüttner distribution (30) for M_p =2.895E-8 kg.

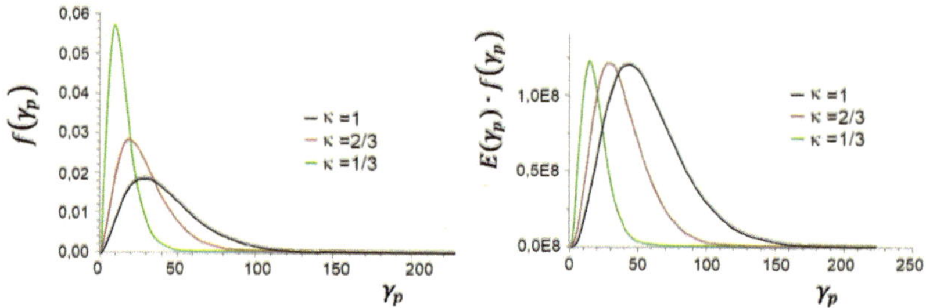

Fig. 10 graphically represents the results of the evaluation of the total energy of the field bosons, integral (35), for test particles of differing heaviness with the linkage constants $\kappa = 1$, 2/3 and 1/3. The integral (35) with $\kappa = 1/3$ therefore corresponds to the energy equivalent of the mass m_T of a test particle (37). See values marked with a cross in Fig. 10.

$$m_T \cdot c^2 \cong \int_{\gamma_p=1}^{\gamma_p=\infty} E(\gamma_p) \cdot f(\gamma_p) \, d\gamma_p \qquad \kappa = 1/3 \tag{37}$$

Figure 10: Comparison of mass-equivalent energy test particles $m_T \cdot c^2$ vs. field boson energies with different κ-values.

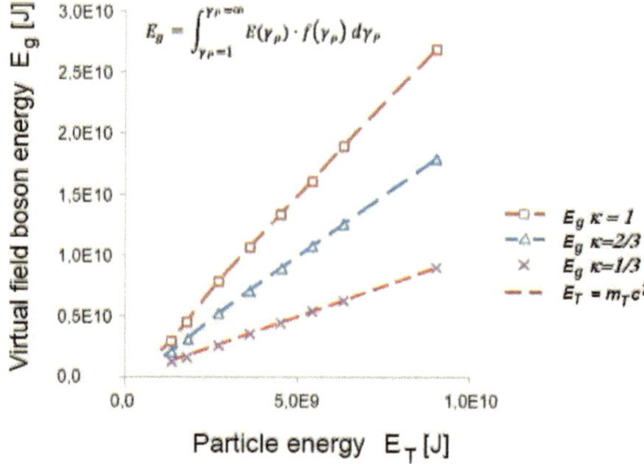

Except for small discrepancies, three times the energy equivalent of a test particle (38) result from the numerical solution of the integral equation (35) by means of the projection masses $M_{p,n}$ with $\kappa = 1$ (Fig. 10, brown curve).

$$3 * m_T \cdot c^2 \cong \int_{\gamma_p=1}^{\gamma_p=\infty} E(\gamma_p) \cdot f(\gamma_p)\, d\gamma_p \,, \qquad \kappa = 1 \tag{38}$$

The integral (35) for $\kappa = {}^2/_3$ results in the blue curve in Fig. 10, which corresponds to double the energy equivalent of a test particle. Table 2 lists the corresponding numerical values.

At a mass scale between 0 and 1,5E-8 kg the velocity trajectories do not pass an energy minimum (fig. 4), so no stable field bosons are to be expected in this area. As a result elementary particles are not perceived by field bosons in these area. However, elementary particles can be "recognized" by field bosons of high scale mass, so they get also mass mediated. In this case the intersection point of the test particle energy with the lower limit of the phase must be used. The energy distribution with a corresponding projection mass $M_{p,n}$ must be rescaled, otherwise it works analogous to a "recognition" by the upper limit of the phase.

Table 2: Numerical values for the test particle energies, projection mass $M_{p,n}$, and the total energy per (35) for the κ-values 1/3, 2/3, and 1 with different test particle energies.

Mass m_T of Test Particle [kg]	Projection Mass $M_{p,n}$ [kg]	Equiv. Energy Test Particle E [J]	$\int_1^\infty f(\gamma_p) \cdot d\gamma_p$	$\int_{\gamma_p=1}^{\gamma_p=\infty} E(\gamma_p) \cdot f(\gamma_p)\, d\gamma_p$ [J] for		
				κ=1	κ=2/3	κ=1/3
1.5E-8	1.055E-08	1.35E+09	1.000	3.001E+09	2.108E+09	1.267E+09
2E-8	1.670E-08	1.80E+09	1.000	4.579E+09	3.109E+09	1.680E+09
3E-8	2.895E-08	2.70E+09	1.000	7.880E+09	5.223E+09	2.620E+09
4E-8	3.956E-08	3.60E+09	1.000	1.067E+10	7.112E+09	3.558E+09
5E-8	4.978E-08	4.49E+09	1.000	1.342E+10	8.947E+09	4.474E+09
6E-8	5.987E-08	5.39E+09	1.000	1.615E+10	1.076E+10	5.381E+09
7E-8	6.992E-08	6.29E+09	1.000	1.898E+10	1.259E+10	6.285E+09
1E-7	1.000E-07	8.99E+09	1.000	2.697E+10	1.798E+10	8.987E+09

5.0 Discussion and Conclusion

The physical properties of field bosons described in this work can perhaps best be paraphrased in terms of the spacetime particle of the Higgs field, to which effective mass and a spatial expansion in the magnitude of the effective de Broglie wavelength can be attributed. The granularity of space-time already begins at energies far below the Planck scale (see Fig. 7 lower right) and depends on the relative velocity of the field bosons as compared to the inertial system. However, according to the SBM, the granularity of space-time cannot be proven by a dispersion of light speed. See also the explanations in the section "the Schwarzschild-de Broglie modification of the SRT for massive field bosons."

In summary, the properties of field bosons can be classified as follows:

- Relativistic mass effects can appear at comparatively low relative velocities of field bosons in comparison to an observer or object (inertial systems)
- In flat space, only a third of the total field boson energy present creates a corresponding equivalent mass

- Two-thirds of field boson energies have no additional effect on an observer or the mass of a test particle and can be linked with dark energy in flat space.

Therefore, two thirds of field boson energies remain, and their origin is completely unexplained. It is conceivable that, during the process of a slow roll inflation, high-energy field bosons lose effective mass through deformation pressure and the gradual condensation of the field bosons. The diminishing effect of gravitation related to this process creates an energy potential, which in connection with a particle could be transferred back to field bosons. The energy of the particle could then contribute to the excitement of the field bosons with a kind of booster effect.

In Fig. 11, a vectored field boson excitation is schematically presented close to a test particle T with its center of gravity in the middle of the sphere. The x-component in the direction of the particle's center of gravity controls the generation of mass. In an asymptotic flat space-time (i.e. without gravitation), field boson excitations are equally probable in all three directions, so that statistically, only 1/3 of the field boson energy can be used towards the creation of particle mass. Thus, the same result is found in both the integral of energy distribution of field bosons per equations (37, 38) and in Fig. 12.

Figure 11: Vectored field boson excitation

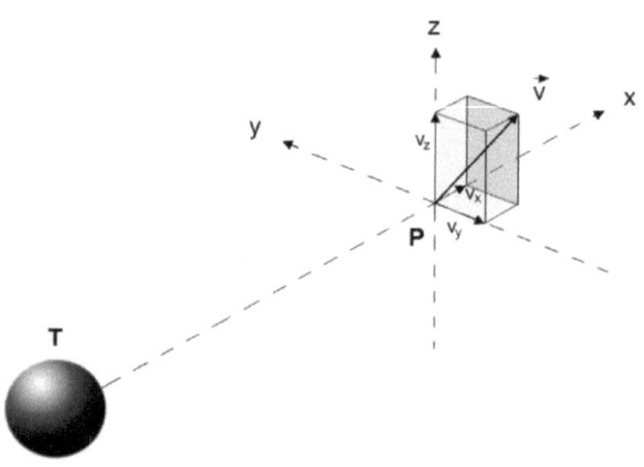

In a scenario with a previously unknown elementary particle, one would thus expect about double the amount of dark energy as dark matter in flat space. Following this idea, one would expect also a greater mass density of dark matter in the center of Galaxies as actually is found.

On the other hand, the SBM model can be interpreted as an alternative to this hypothesis. For an observer or an object, the relative dynamic of (quasi) free field bosons can be made responsible for the uncommon effects. A Higgs field structured after the SBM model, either drifting towards or away from an observer, would show a relativistic increase in mass – even at speeds far below the speed of light. In this sense, the SBM model resolves many inconsistencies regarding the interpretation of dark matter and dark energy in well-known theories.

6.0 Acknowledgement

Special thanks go to my wife Christine, who contributed greatly to the success of this work with her valuable questions and comments.

7.0 Literature

[1] Planck 2013 results. XVI. Cosmological parameters, arXiv:1303.5076, Submitted to Astronomy & Astrophysics.
[2] F. Zwicky (1933), Die Rotverschiebung von extragalaktischen Nebeln, Helvetica Physica Acta,Vol. IV, S. 110 (1933).
[3] F. Zwicky (1937), On the Masses Nebulae and of Clusters of Nebulae, Astrophysical Journal, vol. 86, p. 217 (1937).
[4] Bosma, A (1998), In Galaxy Dynamics, Rutgers University,ASP Conf. Serie 182, S. 339.
[5] Ostriker, J. P. & Caldwell, J. A. R. 1979, in IAU Symposium, Vol. 84, The Large-Scale Characteristics of the Galaxy, ed. W. B. Burton, 441–448.
[6] Schmidt, M. 1985, in IAU Symposium, Vol. 106, The Milky Way Galaxy, ed.H. van Woerden, R. J. Allen, & W. B. Burton, 75–81.
[7] Kent, S. M. 1986, AJ, 91, 1301.
[8] Navarro, J. F., Frenk, C. S., & White, S. D. M. 1997, ApJ, 490, 493.
[9] van Albada, T. S., Bahcall, J. N., Begeman, K., & Sancisi, R. 1985, ApJ, 295,305.
[10] C. Moni Bidin *et al.* 2010 *ApJ* **724** L122 doi:10.1088/2041-8205/724/1/L122The Astrophysical Journal Letters Volume 724 Number 1 .
[11] Bo Qin, Ue-Li Pen, Joseph Silk, http://arxiv.org/abs/astro-ph/0508572.
[12] Martin Bojowald 2007, The Dark Side of a Patchwork Universe http://arxiv.org/abs/0705.4398v1.
[13] Lawrence Krauss: Quintessence - the mystery of missing mass in the universe. Basic Books, New York 2000, ISBN 0-465-03740-2.
[14] Markus Kuster, et al.: Axions - theory, cosmology, and experimental searches. Springer, Berlin 2008, ISBN 978-3-540-73517-5.
[15] Robert R. Caldwell, Marc Kamionkowski, Nevin N. Weinberg (2003)Phantom Energy and Cosmic Doomsday, Phys.Rev.Lett. 91 (2003) 071301.
[16] M. Milgrom, A Modification of the Newtonian Dynamics as a possible Alternative to the Hidden Mass Hypothesis, The Astrophysical Journal, 270: 365-370,1983, Juli 15.
[17] R. Cen, JP Ostriker, - X-ray clusters in a cold dark matter + lambda universe: A direct, large-scale, high-resolution, hydrodynamic simulation,The Astrophysical Journal, 1994 - adsabs.harvard.edu.
[18] N. Jarosik, C. L. Bennett, J. Dunkley, B. Gold, M. R. Greason, M. Halpern, R. S. Hill, G. Hinshaw, A. Kogut, E. Komatsu, D. Larson, M. Limon, S. S. Meyer, M. R. Nolta, N. Odegard, L. Page, K. M. Smith, D. N. Spergel, G. S. Tucker, J. L. Weiland, E. Wollack, E. L. Wright (2010), Seven-Year Wilkinson Microwave Anisotropy Probe (WMAP) Observations: Sky Maps, Systematic Errors, and Basic Results, arXiv:1001.4744 astro-ph.CO].
[19] P. Kroupa et. al., Local-Group tests of dark-matter Concordance Cosmology: Towards a new paradigm for structure formation, arXiv:1006.1647 [astro-ph.CO].
[20] Moore, Ben, et al. "Cold collapse and the core catastrophe." *Monthly Notices of the Royal Astronomical Society* 310.4 (1999): 1147-1152.
[21] Governato, C. Brook, L. Mayer, A. Brooks, G. Rhee, J. Wadsley, P. Jonsson, B. Willman, G. Stinson, T. Quinn & P. Madau: Bulgeless dwarf galaxies and dark matter cores from supernova-driven outflows, Nature 463, 203-206 (14 January 2010).

[22] Benoit Famaey, Stacy McGaugh, Modified Newtonian Dynamics (MOND): Observational Phenomenology and Relativistic Extensions, arXiv:1112.3960 [astro-ph.CO].

[23] B. Famaey, S McGaugh , Challenges for ΛCDM and MOND Journal of Physics: Conference Series, 2013 - iopscience.iop.org.

[24] M. Milgrom, A Modification of the Newtonian Dynamics as a possible Alternative to the Hidden Mass Hypothesis, The Astrophysical Journal, 270: 365-370,1983, Juli 15.

[25] A.P. Lundgren, M. Bondarescu, R. Bondarescu, J. Balakrishna, Astrophys. J. Lett. 715 (2010) L35.

[26] I. Rodriguez-Montoya, J. Magaña, T. Matos, A. Pérez-Lorenzana, Astrophys. J. 721 (2010) 1509.

[27] T.P. Woo, T. Chiueh, Astrophys. J. 697 (2009) 850.

[28] L.A. Ureña-López, JCAP 0901 (2009) 014.

[29] S. Fagnocchi, S. Finazzi, S. Liberati, M. Kormos, A. Trombettoni, New J. Phys. 12 (2010) 095012.

[30] T. Harko, Mon. Not. Roy. Astron. Soc. 413 (2011) 3095.

[31] A. Suárez, T. Matos, arXiv:1101.4039 [gr-qc].

[32] T. Harko, F.S.N. Lobo, arXiv:1104.2674 [gr-qc].

[33] L.A. Gergely, T. Harko, M. Dwornik, G. Kupi, Z. Keresztes, arXiv:1105.0159[gr-qc].

[34] T. Harko, Phys. Rev. D 83 (2011) 123515.

[35] T. Harko, Mon. Not. Roy. Astron. Soc. 413 (2011) 3095.

[36] P.H. Chavanis, Phys. Rev. D 84 (2011) 043531.

[37] J. Barranco, A. Bernal, J.C. Degollado, A. Diez-Tejedor, M. Megevand, M. Alcubierre, D. Núñez, O. Sarbach, arXiv:1108.0931 [gr-qc].

[38] J. Barranco, A. Bernal, arXiv:1108.1208 [astro-ph.CO]

[39] Dan Hooper and Tilman Plehn (2003), Supersymmetric Dark Matter - How Light Can the LSP Be?, Phys.Lett.B562:18-27,2003 (http://arxiv.org/abs/hep-ph/0212226)

[40] Giovanni Amelino-Camelia: Planck scale effects in astrophysics and cosmology. Springer, Berlin 2005, ISBN 3-540-25263-0

[41] M. Planck, Mai 1899, *Über irreversible Strahlungsvorgänge , Sitzungsberichte der Preußischen Akademie der Wissenschaften* (Band 5 S. 479 1899)

[42] Richard L.Amoroso: *Gravitation and cosmology - from the Hubble radius to the Planck scale.* Kluwer Academic, Dordrecht 2002, ISBN 1-4020-0885-6

[43] Amelino-Camelia, G. (2010). "Doubly-Special Relativity: Facts, Myths and Some Key Open Issues". *Symmetry* 2: 230–271. arXiv:1003.3942. Bibcode:2010arXiv1003.3942A.

[44] Magueijo, J.; Smolin, L (2001). "Lorentz invariance with an invariant energy scale". *Physical Review Letters* **88** (19): 190403. arXiv:hep-th/0112090. Bibcode:2002PhRvL..88s0403M. doi:10.1103/PhysRevLett.88.190403.

[45] Magueijo, J.; Smolin, L (2003). "Generalized Lorentz invariance with an invariant energy scale". *Physical Review D* **67** (4): 044017. arXiv:gr-qc/0207085. Bibcode:2003PhRvD..67d4017M. doi:10.1103/PhysRevD.67.044017.

[46] A. A. Abdo et al. A limit on the variation of the speed of light arising from quantum gravity effects, *Nature* **462**, 331-334 (19 November 2009) | doi:10.1038/nature08574; Received 12 August 2009; Accepted 12 October 2009; Published online 28 October 2009

[47] P. Laurent, D. Götz, P. Binétruy, S. Covino, A. Fernandez-Soto. Constraints on Lorentz Invariance Violation using integral/IBIS observations of GRB041219A. *Physical Review D*, 2011; 83 (12) DOI: 10.1103/PhysRevD.83.121301

[48] Linde, A (1982). "A new inflationary universe scenario: A possible solution of the horizon, flatness, homogeneity, isotropy and primordial monopole problems". *Physics Letters B* **108** (6): 389–393. Bibcode:1982PhLB..108..389L. doi:10.1016/0370-2693(82)91219-9.

[49] Albrecht, Andreas; Steinhardt, Paul (1982). "Cosmology for Grand Unified Theories with Radiatively Induced Symmetry Breaking" (PDF). *Physical Review Letters* **48** (17): 1220–1223.

[50] W. J. G. de Blok, "The Core-Cusp Problem," Advances in Astronomy, vol. 2010, Article ID 789293, 14 pages, 2010. doi:10.1155/2010/789293.